24th Topical Conference on Refinery Processing 2021

Topical Conference at the 2021 AIChE Spring Meeting and 17th Global Congress on Process Safety

Online
18 – 23 April 2021

ISBN: 978-1-7138-5332-9

Printed from e-media with permission by:

Curran Associates, Inc.
57 Morehouse Lane
Red Hook, NY 12571

Some format issues inherent in the e-media version may also appear in this print version.

Printed with permission by Curran Associates, Inc. (2022)

For permission requests, please contact AIChE
at the address below.

AIChE
120 Wall Street, FL 23
New York, NY 10005-4020

Phone: (800) 242-4363
Fax: (203) 775-5177

www.aiche.org

Additional copies of this publication are available from:

Curran Associates, Inc.
57 Morehouse Lane
Red Hook, NY 12571 USA
Phone: 845-758-0400
Fax: 845-758-2633
Email: curran@proceedings.com
Web: www.proceedings.com

TABLE OF CONTENTS

619115 Application-Oriented Review of Data Reconciliation and Gross Error Detection Methods for Heat Exchanger Networks

Abstract:

Application-oriented Review of Data Reconciliation and Gross Error Detection Methods for Heat Exchanger Networks

José Loyola-Fuentes, Carlos Planelles Alemany, Emilio Diaz-Bejarano and Francesco Coletti*

Hexxcell Ltd., Foundry Building, 77 Fulham Palace Rd, London W6 8AF, UK

*corresponding author: f.coletti@hexxcell.com

Extracting useful insights from raw plant data is often a challenging and frustrating experience. On one hand, measurements may be subject to random errors (e.g. arbitrary fluctuations in the environment, signal transmission) and/or systematic errors (measurement bias, instrument degradation, data corruption or complete sensor failure). On the other hand, plant operators rely on these measurements (e.g. flow rates, temperatures, pressures) for process and asset monitoring and make important operating and maintenance decisions based on this information. It is therefore clear that rigorous approaches capable of dealing with the various sources of error are needed.

The field of statistics provides a substantial number of filtering techniques that are able to identify significant disturbances in the measured data, but the physical context where the process states are embedded in is not entirely considered. Therefore, a suitable method that combines the physical and statistical knowledge of a specific system could be beneficial for monitoring purposes. Data reconciliation (DR) and gross error detection (GED) are complementary methods that reduce the effect of random and gross errors by exploiting the relationship between measured variables and a specific process model. The estimations of process variables via DR are expected to be unbiased, more accurate and in accordance with the process model, which in case of heat exchanger networks, could be mass and energy conservation equations. The role of GED is to deliver one or more of the following objectives: to detect the presence of gross errors, to find the sources of single/multiple gross error and to estimate the magnitude of such errors. The challenges related to the completion of these tasks (including DR) depend on the conditions and assumptions associated with the process model, and throughout the past 50 years, a wide variety of methods have been developed to address different data reconciliation problems.

In this presentation, we critically review current practice and state-of-the-art in data reconciliation and gross error detection techniques, applied to heat exchanger networks. The review includes considerations for both steady-state and dynamic systems and an application-oriented summary of the advantages and disadvantages of various techniques (e.g. classic and robust statistics, Kalman filtering, dynamic programming) and presents relevant case studies.

619320 Refinery-Petrochemical Integration – Repurpose Assets, Maximize Profitability, and Pivot to the Future

Abstract:

The refining industry is undergoing a monumental shift as demand for fuels is projected to wane, while demand for petrochemicals continues to rise. To remain viable, refiners will need to repurpose existing assets, upgrade low-value streams, and pivot towards petrochemicals production.

There is no 'one-size fits all' solution to refinery-petrochemicals integration. The optimal solution for each site will vary and refiners must consider several key factors including regional dynamics and infrastructure availability, and which value chain to pursue. Maximization of existing assets is paramount to keep project costs down and drive economics.

There are also different levels of integration. Basic refinery grade petrochemicals are common nowadays, but higher levels of integration include production of high purity products and adding traditional petrochemicals units such as steam cracking or on-purpose olefins. The fullest extent of integration can be realized in a full crude to chemicals (COTC) complex where conversion levels to both fuels and chemicals can exceed 90%.

KBR has a broad portfolio of integration technologies including ROSE® solvent deasphalting, SCORE™ steam cracking, K-COT™ catalytic olefins technology, K-PRO™ propane dehydrogenation, VCC™ slurry hydrocracking, AED-BTX aromatics extraction, and MAXOFIN™ maximum olefins FCC. All these technologies are enablers to shift from fuels to petrochemicals production.

KBR explores common regional refinery configurations and practical solutions. The path to the future and the right integration solution can be developed through experience and a detailed analysis of existing refinery configurations.

618751 Optimize Gasoline Block Using Dividing Wall Column Technology

Abstract:

Gasoline Block in refinery processes full range naphtha (C5 to C12 carbon atoms) to useful fractions. These fractions are primarily used for gasoline blending or as feedstock to petrochemical complexes.

For gasoline blending, improvement in RON is of prime importance. A Naphtha splitter splits full range naphtha (FRN) into light naphtha (C5/C6) and heavy naphtha (C7+). Increase in RON for light Naphtha (C5/C6) is made possible by converting straight chain paraffins into their branched chain isomers through the process of isomerization. Heavy Naphtha (C7+) RON is improved through the process of catalytic reforming. Heavy Naphtha (C8+) can also be used as feedstock to paraxylene complex (PX).

Naphtha splitter pays a pivotal role in feed preparation to Isom, CCR and PX Complexes. Dividing wall column technology (DWC) is increasingly used to debottleneck conventional distillation columns. Besides energy and throughput increase, DWC revamp of naphtha splitters significantly improve the performance of downstream units. This presentation explores four case studies in revamp of gasoline block:

- Revamp of a **sequence of naphtha splitters to DWC** preparing feed for Isom and CCR Units.
- Revamp of a **Side cut naphtha splitter to DWC** preparing feed for Isom and CCR Units.
- Revamp of a **Side cut naphtha splitter to DWC** preparing feed for Isom and CCR & PX Units.
- Revamp of a **reformate splitter** to offload down stream aromatic extraction unit.

In all the above scenarios the revamp to DWC offered a payback of few months to less than a year.

619330 Techno-Economic Analysis of Direct Methane Aromatization for Natural Gas Upgrading at Isolated Sites

Abstract:

Inexpensive sources of natural gas have led to a boom in development of technologies taking advantage of C1 chemistry; however, a significant number of potential natural gas sources are stranded and lack the infrastructure for economically-feasible transportation. The resulting gas, which is normally flared, can, in theory, be recovered at the well-site, purified, and then processed to a product which is easier to transport. Gas-to-liquid (GTL) processes exist which can convert the feed to liquid products which can be more easily transported, although there are process constraints due to the need for deployment in remote locations, and several are being commercialized. [1] One such process, direct methane aromatization (DMA), produces significant amounts of benzene, a valuable commodity chemical, and hydrogen. The benzene can be purified on-site and then transported by truck; the hydrogen can be dealt with on-site as well by either transforming it to another transportable chemical, such as ammonia or cyclohexane, or using it to provide heating or electricity.

Previous economic studies of DMA have been reported which either invoke unrealistic assumptions in the process model, [2,3] do not consider utilization of the gaseous hydrogen, [2,3] or do not consider the limitations imposed on the process due to remote deployment. [4] With these issues in mind, a DMA mini-plant processing 1.0 MMSCFD of natural gas was designed with currently available technology to provide a platform for the investigation of process intensified alternatives. The reactor section of the plant uses a pre-carburized DMA catalyst developed at Texas Tech. [5] The unit was investigated for economic feasibility and used to draw conclusions about the current state of DMA technology. In particular, the analysis identifies five particular areas which can benefit from intensification: aromatics yield (i.e. one-pass conversion), coke selectivity, hydrogen recovery, and separation and purification of both the aromatic products and generated hydrogen. Although intensified reactor technology for the DMA process is already being investigated for improving the yield with [6,7] and without integrated hydrogen recovery, [8, 9] the process model framework should provide opportunities for investigating additional intensification technologies, particularly in separations.

1. H. Fleisch., T. [White Paper], World Bank-Global Gas Flaring Reduction Partnership, 2015.
2. Pérez-Uresti, S.I. *et al.*, *Processes*, 2017, 5, 33.
3. Camilo Corredor, E., P. Chitta, and M. D. Deo. *Fuel Pro. Tech.*, 2019, 183, 55 – 61.
4. Huang, K. *et al.*, *Joule*, 2018, 2, 349 – 365.
5. Rahman, M., A. Sridhar, and S.J. Khatib. *Cat. A Gen.*, 2018, 558, 67 – 80.
6. Kee, B. *et al.*, *Eng. Chem. Res.*, 2017, 56, 3551 – 3559.
7. Morejundo, S.H. *et al.*, *Science*, 2016, 353(6299), 563 – 566.
8. Brady, C., B. Murphy, and B. Xu. *ACS Catal.*, 2017, 7(6), 3924 – 3928.
9. Cao, Z. *et al.*, *Chem. Int. Ed.*, 2013, 52, 13794 – 13797.

618581 Development of an Integrated Process Plant for the Conversion of Shale Gas to 1,3 Butadiene

Abstract:

1,3 butadiene is an important feedstock in the production of rubbers and plastics, such as styrene butadiene rubber, polybutadiene rubber, and styrene butadiene latex. As the cracker feedstock around the globe is trending towards lighter feedstock from shale gas, the sustained production of 1,3-butadiene in olefin plants, which is traditionally made via naphtha cracking, is facing big challenges. In this research, a novel integrated plant is developed for the production of 1,3 butadiene from shale gas. The manufacturing process consists of the following steps: (1) conversion of shale gas to methanol, (2) conversion of methanol to ethylene and (3) conversion of ethylene to 1,3 butadiene. In the first step, shale gas is used to produce syngas via steam reforming, which is subsequently transformed to methanol. In the second step, methanol is catalytically converted to ethylene via the MTO process. In the third step, ethylene undergoes a dimerization reaction to produce 1,3 butadiene. The conversion of ethylene is significantly impacted by temperature and the 1,3 butadiene mole fraction in the product stream can vary from about 21% to 31% in the temperature range 1273K – 3273K. Currently, each of these steps are conducted in independent process plants and so there are opportunities to utilize process intensification tools in the integrated process plant. The overall plant is simulated in the ASPENPlus environment and the predictive capabilities of this model is tested by comparing the results from experimental data from individual plants. Then, a variety of process conditions are tested at steady state to optimize the production of 1,3 butadiene. Heat-integration tools are utilized for energy-saving and capital cost reduction opportunities. A comparative economic assessment based on existing plant information indicates that the use of process integration techniques has the potential to reduce costs significantly.

NOTES

NOTES

NOTES

NOTES

AUTHOR INDEX